EVERY GARDEN
IS A STORY

EVERY GARDEN IS A STORY

Stories, Crafts, and Comforts

Susannah Seton

Foreword by Carolyn Rapp

Conari Press

First published in 2007 by Conari Press,
an imprint of Red Wheel/Weiser, LLC
With offices at:
500 Third Street, Suite 230
San Francisco, CA 94107
www.redwheelweiser.com

Much of the material in this collection was previously published in *Simple Pleasures of the Garden* by Susannah Seton, Berkeley, CA: Conari Press, 1998. ISBN:1-57324-104-0 (hardcover); ISBN: 1-57324-501-1 (paperback).

ISBN-10: 1-57324-318-3
ISBN-13: 978-1-57324-318-6
Library of Congress Cataloging-in-Publication Data
Seton, Susannah, 1952–
  Every garden is a story : stories, crafts, and comforts /Susannah Seton.
    p. cm.
  ISBN 1-57324-318-3 (alk. paper)
  1. Gardening—Anecdotes. 2. Gardens—Anecdotes. 3. Nature craft. 4. Recipes. I. Title.
  SB455.S429 2007
  635--dc22

                    2007011601

Cover and text design by Donna Linden
Typeset in Futura and Perpetua

Cover photograph © Alison Miksch/Brand X Pictures, Interior photographs: Pages 7, 12, 15, 20, 25, 27, 37, 41, 50, 65, 67, 79, 85, 99, 110, 115, 121, 123, 125, and 126 © Alison Miksch/Brand X Pictures; pages 2, 6, 19, 30, 117, and 128 © Corbis; page 8 © Tomo Jesenicnik/iStockphoto; page 22 © Ewa Kubicka/iStockphoto; page 35 © Jean Schweltzer/iStockphoto; page 39 © Sherry Holub/iStockphoto; page 42 © Robyn Mackenzie/iStockphoto; page 45 © smiley joanne/iStockphoto; pages 47, 53, 57, 75, 92, and100 © iStockphoto; page 49 © Christine Glade/iStockphoto; page 55 © Don Enright/iStockphoto; page 60 © Vladimir Ivanov/iStockphoto; page 63 © Stuart Berman/iStockphoto; page 69 © Oleg Kazlov/iStockphoto; page 70 © Massimiliano Pieraccini/iStockphoto; page 73 © Karin Lau/iStockphoto; page 77 © Tom Ediger/iStockphoto; page 80 © Joseph Justice/iStockphoto; page 83 © Annett Vauteck/iStockphoto; page 89 © Stephanie Daoust/iStockphoto; page 90 © Frank Richard Kebschull/iStockphoto; page 95 © Pamela Moore/iStockphoto; page 105 © Roger McLean/iStockphoto; page 109 © Andrew Howe/iStockphoto; page 113 © Jon Tyler/iStockphoto; page 118 © Peggy Easterly/iStockphoto.

Printed in China
MD
10 9 8 7 6 5 4 3 2 1

A book is like a garden carried in the pocket.
CHINESE PROVERB

CONTENTS

FOREWORD

When I was a child, I had the good fortune of growing up next door to my grandmother, an avid gardener who always had something wonderful in bloom. Each year in late May, the star of the garden was Nana's climbing roses—dazzling cherry-red blooms so abundant that they covered the side of her house and had to be tied to the trellis to save them from their own joyful profusion. As soon as there were enough, Nana would come fetch me, scissors in hand, to cut the first bouquet. That one was for me to take to my teacher. Dozens more followed – for neighbors up and down the street, for the church, for friends at the nursing home, for my mother's kitchen table, for the counter at the corner grocery store. Often I was the delivery girl, so I got to see firsthand what these gifts from the garden meant to those who received them—from surprise to delight to tears. From Nana's yearly "ritual of the roses," I learned early on a truth about gardeners: they are generous souls.

No wonder I fell immediately in love with *Every Garden Is a Story!* In these pages, Susannah Seton offers her readers a bouquet as generous and abundant as any I delivered for Nana. And like my grandmother's roses, this bouquet of stories, garden tips, recipes, quotes, and clever how-to ideas surprises, intrigues, delights, and touches the heart.

The stories that Susannah has so carefully chosen take us into gardens from the northeast coast of England, with its harsh climate and stony soil, to the blistering sun and brassy blue skies of California. But, as you'll discover, where the garden grows or what is grown in it matter little. It is the passion for gardening and the openness to receiving Earth's lessons that create a common language among gardeners, no matter where they turn the soil. I found kindred spirits on every page of the book.

Who could not be inspired by the man with a tiny yard and enormous optimism who, against all odds, made his dream of creating a seven-headed sweet pea come true? Who among us doesn't have a dream? As for weeding, I understand perfectly how satisfying the "weeding vacation" was to the woman who, like so many of us in the modern workplace, is immersed in the intangible world of ideas. She transformed weeding from drudgery into meditation and took pleasure in a kind of work that marks tangible progress.

Some of the most touching stories remind us that you don't have to have a big yard or a lot of money to have a garden. A sweet potato propped with toothpicks in a Mason jar soon filled the kitchen window of a tiny three-room apartment with a curtain of graceful vines, transforming a shabby little room into a place of wonder. The magic of that sweet potato also awakened in a small child a lifelong love of gardening.

Mixed in with the stories is a multitude of practical tips and fun how-tos. How to attract butterflies to your garden and how to repel pests without poisoning your soil with chemicals. Where to purchase an army of ladybugs to combat aphids on your roses. How to cut a bouquet to make it last the longest. How to dry hydrangeas properly, decorate serving platters with rosemary rings, grow produce if you live in an apartment, beautify your old straw hat with a ribbon of flowers, and grow your own loofah sponges (did you know they are actually gourds?).

One of my favorites is creating a memory garden to honor friends and family. Give a rhododendron as a sympathy gift. Plant one in your own garden to remind you of a friend or relative that has passed away or moved away. Plant a "birth tree" for a child, then involve that child in helping to take care of the tree and track its growth by tying a bit of yarn to the out most tip of a branch each fall and seeing where the yarn ends up after the next summer. Swap seeds and bulbs to create a "friendship" garden that reminds you of your connections as you work or walk in it.

And when you come in from the garden, exhausted and covered with dirt, this lovely collection offers recipes to soothe your aches and reward you for your hard work. You can start by relaxing in the tub with bath oil made from lavender grown in your garden, then pamper your skin with the peaches and cream skin moisturizer, then treat yourself to a cup of tea, a delicious blend you can make with basil, mint, and rose hips from the garden.

My friend and master storyteller Carol Birch says that stories undo the damage of haste, and *Every Garden Is a Story* is sure to slow you down. It is a book for all seasons, delighting in the full bloom of the garden and also encouraging us to see beauty in its winter garb, when the grace of stem and branch create exquisite patterns visible only as the earth rests and regenerates. It is a book that stimulates all the senses— don't forget to breathe deeply the scents of these gardens! It is a book that you can pick up and dive into anywhere—start your day with a story to inspire and end it with some of the stunningly beautiful photos.

But what I like best is the way this book keeps giving and giving. Each story I read evoked a story of my own, connecting me with special moments and special people in my life. The miracle of seeing my first lettuce sprout from those infinitesimal seeds. The joy of friendships grown in a community garden. Making puppets out of snapdragons and ladies in ball gowns out of the hollyhocks that grew by the back alley fence. A sweet-scented gardenia brought by a friend when I was sick. Giving away my first bouquet of zinnias from my first garden— with a generous heart, just like Nana and her bouquets of roses.

You are about to embark on a lovely journey. Begin —and see where the stories take you!

—CAROLYN RAPP, author of *Garden Voices: Stories of Women
& Their Gardens*

EVERY GARDEN IS A STORY

DOWN THE GARDEN PATH

Whenever I arrive in my garden, I "make the tour." By "making the tour,"
I mean only that I step from the front window, turn to the right, and make
an infinitely detailed examination of every foot of ground, every shrub and
tree, walking always over an appointed course.

There are certain very definite rules to be observed when you are making the tour. The chief rule is that you must never take anything out of its
order. You may be longing to see if a crocus has come out in the orchard, but
it is strictly forbidden to look before you have inspected all the various beds,
bushes, and trees that lead up to the orchard.

You must not look at the bed ahead before you have finished with
the bed immediately in front of you. You may see, out of the corner
of your eye, a gleam of strange and unsuspected scarlet in the
next bed, but you must steel yourself against rushing to this
exciting blaze, and you must stare with cool eyes at the earth in
front, which is apparently blank, until you have made certain
that it is not hiding anything. Otherwise, you will find that you
rush wildly round the garden, discover one or two sensational
events, and then decide that nothing else has happened.

*One of the most
delightful things
about a garden is
the anticipation
it provides.*
W. E. JOHNS

INNER PEACE

Working in the garden gives me a profound feeling of inner peace. Nothing here is
in a hurry. There is no rush toward accomplishment, no blowing of trumpets. Here is the
great mystery of life and growth. Everything is changing, growing, aiming at something,
but silently, unboastfully, taking its time.

HIS SOUL WALKED WITH FLOWERS

My father was an ambitious man, though not in the usual way. His garden of beautiful roses was important to him, but most of all, he wanted, with all of his heart, to grow a seven-headed sweet pea. To his knowledge, this had never been done before.

The harsh climate and stony soil of the northeast coast of England was more conducive to the production of stalwart vegetables, coarse in quality but gargantuan in size. Local gardeners competed in shows for the biggest and best. My father was a different kind of dreamer. His soul walked with flowers.

We lived in a dirty, decaying city grappling with the decline of shipbuilding and coal mining. Our small red brick house was skirted on three sides by the tiniest of gardens, its perimeters defined by the ubiquitous privet hedge. Reluctant sun struggled through the constant, dour overcast of clouds and was mostly blocked by nearby buildings. Did he not realize that roses needed sunshine? Then there were the cutting arctic winds, icy as snow stars that keened year round through every crevice. They blew the hope out of such fragile blooms as sweet peas. As I said, he was a dreamer, utterly convinced that a seven-headed flower could be coaxed from his plants, not by the technicalities of cross-fertilization, but by sheer love and devotion.

Each flower is a soul opening out to nature.
GÉRARD DE NERVAL

Our neighbors—if they had gardens at all—had geometric patches of grass bordered by regimental rows of marigolds and snapdragons. His was a rock garden with cascading lobelia and alyssum, sheltered by a richly hued copper birch tree. Then there was the rose bed, each plant named with pride of place, leading to his favorite, the exotic, heavily scented Crimson Glory. And there was the bed of sweet peas, each plant supported by elaborate scaffolding. Everything flourished.

My father held his ambition lightly. It was the process of gardening that delighted him. With the diligence and dedication of an acolyte, he was out daily, fumbling with loving but clumsy hands at the delicate tendrils of the sweet peas as he cajoled them into climbing further up their supporting frames.

He nourished them with more than love alone. The manure heap sat by the kitchen door. It was supplemented by steaming pails of fresh droppings from the horse-drawn produce, fish, and coal carts that delivered to our street. Fly papers flapped from the kitchen ceiling. Mother yelled to keep the door shut.

Inevitably, it happened. One day we were all called out to see the pale lavender, gossamer, seven-headed sweet pea. Its photograph was ceremoniously taken. With utmost tenderness, my father brought it into the house, surrounding it with maidenhair fern and reverently placing it in a vase at the center of the table. Never have I seen a man so blissful, so content.

I've had many gardens, big and small. Now, in my old age, I have only a tiny deck. It's overflowing with an abundance of flowers, all a little out of control. The blistering sun and brassy blue skies of California cruelly discourage sweet peas. But there are the roses, masses of them in pots, among them the exotic, sweetly scented Crimson Glory. Now there's a memory.

FINDING OLD ROSES

Old roses are not only beautiful, but they have the advantage of being more disease resistant and drought tolerant than the finicky hybrid tea roses.

MOON RIVER VINES

As children we often spent summers with family in the South, summers filled with shade trees, brazenly colored flowers, and choking vegetation in the form of wild morning glories and patches of kudzu that resembled small green seas. With so much growing and tangling about you, it's easy to understand why my family and I love gardening so much. My Uncle Robert loved vines and cultivated, catalogued, and saved seeds, which he doled out to the rest of us. His prize was the moon river vine whose rather ordinary looking seeds held within their brown casing nothing short of a miracle to my cousins and me. Early each summer, Uncle Robert would take the seeds out of an old pill bottle and give us each 3–5 of them with instructions to plant them near a trellis or porch column and keep them watered. Within six weeks, there would be a delicate vine with papery thin leaves that grew at an astonishing rate. By mid-July, the buds would appear luminous white, like twisted-up tissue paper.

The Snowdrop is the
prophet of the flowers;
It lives and ides upon
its bed of snows;
And like a thought of spring
it comes and goes.
GEORGE MEREDITH

From then on, every summer night at dusk we'd gather on the porch with our iced tea and try to act blasé as we furtively glanced every few minutes at the buds of the vine. Our childish patience would eventually be rewarded with a small sway from the vine, sometimes a tiny twitch, then ever so slowly a bud would begin to open before our very eyes, eventually, over the course of about half an hour, unfurling into a saucer-sized pure white flower. The smell was heavenly—magnolias and jasmine—but fainter and more elusive. We were never allowed to touch the flowers or—God forbid—pick them, and I honestly don't think I ever wanted to. I was awestruck by their amazing aliveness. Sometimes we'd turn off the porch light and get to stay up a little later than usual to see if a lunar moth would visit the irresistible flowers; sometimes even a tiny bat would appear. Beyond the porch the lightning bugs would signal to one another, and a symphony of night sounds serenaded us. It was magic.

OF MOON GARDENS AND OTHER THEMES

A moon garden is one that has only white flowers. It's particularly striking at the moment of summer twilight when all colors except white fade and anything white takes on a luminescence that is breathtaking. That makes it a wonderful spot for evening entertaining and for those of us who work so late we only ever see our gardens in the moonlight.

IN AUNT RUTH'S GARDEN

All my hurts
My garden spade can heal.
RALPH WALDO
EMERSON

As a child, I was fortunate enough to have an aunt who loved to garden. She had a beautiful old-fashioned garden in the Ohio valley with many plants that had been handed down through generations of my family. I can still smell the lilac from a bush that was over one hundred years old and home too many bees. When I was barely old enough to toddle along behind her, Aunt Ruth would show me her tricks of the trade—rooting, transplanting, deadheading, even grafting.

I quickly learned that, by admiring one of my aunt's flowers heartily, I would usually be given a seedling, a cutting, or a whole plant. In her infinite wisdom, she decided that I could be taught the craft of gardening and help her clean and thin some beds at the same time. This became a marvelous arrangement that persisted well into my adulthood until I moved 3,000 miles away to a different "zone." Now, as a city dweller, my dream is to have a little garden patch of my own where I can carry on the family tradition.

STEM CUTTINGS

Simply cut off a stem, strip off lower leaves, and place in water. (Don't touch the surface of the cut—bacteria from your skin can make the stem rot.) In a month or so, it will have rooted and you can transplant it into a pot. Avoid rot by placing a couple of charcoal chips at the bottom of the container.

ODE TO ASPARAGUS

The ancients believe that asparagus was an aphrodisiac. It certainly tastes good enough to be, but even if it isn't, the prospect of your own tender shoots each spring should entice you enough to give it a try in your garden. It takes about three years to get enough asparagus to make planting it worthwhile, but a maintained asparagus bed will last for twenty years, so you'll get plenty of spears for your efforts. While asparagus prefers cold winters, it will grow just about anywhere in the United States. The trick is to dig a one-foot-deep trench and half fill it with compost and $\frac{1}{4}$ cup bone meal per one foot of trench. Plant roots 18 inches apart, and don't fill in the trench with dirt until roots begin to sprout.

STIR-FRIED ASPARAGUS

Once you've grown them and had your fill of steamed, give this a try.
It is delicious!

    2 tablespoons low-sodium soy sauce
    1 tablespoon dry sherry
    1 tablespoon water or chicken broth
    1 tablespoon sesame oil
    1½ pounds asparagus, ends snapped, and cut into small
        pieces
    2 teaspoons minced garlic
    2 teaspoons minced fresh ginger
    ½ cup minced fresh basil
    ½ teaspoon sugar

Combine the soy sauce, sherry, and water or broth and set aside.
Place a large (at least 12-inch) skillet over high heat for 4 minutes.
Add 2 teaspoons of the oil and heat for 1 minute or until the oil just
starts to smoke. Add the asparagus and stir-fry for about 2 minutes or
until barely tender. Clear the center of the skillet, add the garlic, ginger, and 1 teaspoon of oil, sauté for 10 seconds. Remove skillet from
heat and stir the ingredients to combine.

   Place skillet back on heat, stir in the soy sauce mixture, and cook
for 30 seconds. Add basil and sugar and cook for another 30 seconds. Serves 4.

Every thing is good in its season.
ITALIAN PROVERB

HEARTFELT GREENS

One easy way to entice your child into the garden is to make a patch
with their name. Simply trace out his or her name in the loose

Green fingers   soil and trace a big heart around it. Then plant a variety of
are the extensions of   fast-growing greens (leaf lettuce, radishes, watercress, aru-
a verdant heart.   gula) in the furrows made by your tracings, water, and wait
RUSSELL PAGE   for his or her name and a big green heart to appear. Chances
are your little one will not only enjoy helping, but he or she
will eat salad too!

FRESH HERB SALAD

This is a salad from Provence that uses an unusual variety of late-spring greens and herbs.

1 garlic clove, peeled and halved
2 teaspoons lemon juice
1/4 teaspoon salt
4 teaspoons olive oil
1 teaspoon hot water
4 cups arugula
1–2 cups watercress, large stems
    removed

1 1/2 cups escarole
1/4 cup parsley, stems removed
1/2 cup curly endive
1/4 cup small basil leaves
20 small tarragon leaves
10 small sage leaves
5 chives, minced
Pepper

Rub the garlic clove halves all over the inside of a large wooden salad bowl. Whisk in the lemon, salt, olive oil, and water. Add greens and pepper to taste. Serve immediately. Serves 4.

SPEAK THE LANGUAGE OF FLOWERS

We all know roses mean "I love you," but do you know the nonverbal messages of other flowers? The Victorians used to practice the "language of flowers," in which they would send little nosegays of homegrown flowers that were actually nonverbal poems. A bouquet of coreopsis and ivy, for example, would mean, "Always cheerful friendship." Such floral messages are called tussy-mussies, a term that dates back to the 1400s when these nosegays first came into fashion; they were routinely carried by both men and women.

The language of flowers was quite complex. If, for example, the flowers were presented upside down, the message was the opposite— an upside down rose, for example, meant "I don't love you."

Flowers preach to us if we will hear.
CHRISTINA ROSSETTI

If the bow or the flower bent to the left, the message referred to the receiver "you have beautiful eyes." If it bends to the right the message refers to the sender. "I send loving thoughts" says the left-leaning pansy. If you added leaves to your tussy-mussy, they signaled hope, while thorns meant danger. When you received a tussy-mussy, touching it to your lips meant you agreed with the message's sentiments. If you tore off the petals and threw them down, you were sending a strong rejection of the sentiment.

Tussy-mussies are easy to make. Simply decide on your message and pick the appropriate flowers, leaving six inches of stem. Strip stems of leaves, and pick the largest flower for the center. Wrap its stem with floral tape. Then add the remaining flowers in a circle, taping the stems together as you go, keeping the height even, until you reach a diameter of about 5 inches. Then add greens, if any. Finish off by winding tape down the length of the stems, and add a ribbon streamer or a piece of lace as a bow. Voila! Send with a card explaining the flowers' meanings. Tussy-mussies should be kept in water.

SAY IT WITH FLOWERS

Apple Blossom: Preference
Azalea: First love
Coreopsis: Always cheerful
Cornflowers: Healing, felicity
Daffodil: Regard
Red Dianthus: Lively and pure affection
Heliotrope: Accommodating disposition
Ivy: Friendship
Johnny-jump-up: Happy thoughts
Lamb's Ear: Gentleness
Lavender: Devotion
Lily of the Valley: Return of happiness
Love-in-a-mist: Kiss me twice before I rise

Mint: Warmth of feeling
Oregano: Joy
Pansy: Loving thoughts
Red Salvia: Energy and esteem
Rosemary: Devotion
Scented Geranium: Preference
Thyme: Courage and strength
Violets: Faithfulness
Wallflower: Fidelity in adversity
White Clover: Good luck
Yarrow: Health
Zinnia: Thought of absent friends

THE BRIDAL WREATH

There was a sadly neglected corner of my parents' yard that had an ancient bedraggled shrub my mother called the bridal wreath. I thought this eyesore was undeserving of such a poetic name since it hardly produced even a handful of white buds and seemed to have a blackberry bush rapidly overtaking it. I had recently learned the fine art of pruning and decided one late fall afternoon to try it on this tired old shrub. I really went at it with the shears and, with the help of my father, got rid of the blackberry invader and hacked away until not much was left of the venerable old bridal wreath. I remember my mother nearly cried when she saw the havoc we had wrought; her mother had planted this plant fifty years before. I told her not to worry and to wait and see what happened.

The garden is a love song, a duet between a human being and Mother Nature.
JEFF COX

It sat there all fall and winter, pretty much forgotten. Then, one early spring day, we all had the most wonderful surprise—the bridal wreath had turned into a showpiece overnight with dozens of sprays of perfect white flowers that lasted for weeks. My mother was happiest of all. Evidently, the bridal wreath had been her mother's pride and joy and was now fully restored to its former glory. It was amazingly beautiful. The sprays were indeed perfect for a bride's crown.

MOTHER'S DAY SACHETS

Here's something lovely to make mom or grandma for Mother's Day.

1/4 yard lace
Dinner plate
Disappearing ink marker
Scissors
Cereal bowl
Tapestry needle
2 yards 1/4-inch-wide wired ribbon
2 ounces lavender or potpourri
2 yards 1-inch-wide ribbon

Place the lace on a table and lay the dinner plate on top of it. Trace the edge of the plate with the disappearing ink marker. Remove plate and cut around marker to make a circle of lace. Turn the cereal bowl upside down in the center of the lace circle and trace the edge. Remove bowl.

Thread the tapestry needle with the thin ribbon and stitch around the inner circle you have just created. (Size of stitches doesn't matter.) Tug gently on the ribbon so the lace gathers to make a pocket. When the opening is the size of a silver dollar, pour the lavender or potpourri in until full (about the size of a walnut). Tug the ribbon tight, tie in a knot, and cut the ends short.

Tie the wide ribbon into a beautiful bow. Repeat until materials are gone. Makes 5 sachets.

MAY, HOME

Living with cancer myself, my healing lessons seem to come now from the garden. No small surprise since it is the place where I find culture and nature meeting. I love this land by the lake fiercely. I am forced to find sanity in the earth. Whenever I get back from traveling, I drop my suitcases at the door, strip off my shoes, and go directly to the dirt.

Yesterday I spent almost two hours being instructed by a huge Siberian iris plant that didn't bloom last year. I guessed it should be divided—-a task I'd never attempted before. Indeed, as I dug in with a pitchfork, I saw that the roots were so entangled, so thickly enmeshed; I had to use a hacksaw to even begin to separate them all. It felt exactly like my mind, like my life. The iris plant has survived fire engines, two dogs' digging, and extreme frost. But what it told me is that its roots got so enmeshed because it didn't have the space it needed to go as deeply as is necessary to bloom. It will take me days to separate all of this complexity and then to build a new raised bed with very deep soil where each shoot can have the room it needs to root deeply enough so it can blossom.

That feels exactly what I need to do with my life as well.

Life begins the day you start a garden.
CHINESE PROVERB

DIVISION

You just take the plant out of the pot, divide it into two or more clumps, roots and all, and replant into two or more pots.

THE TAO OF GARDENING

At one point in junior high, I became such an avid would-be gardener that I worked at a greenhouse during the summer and fancied becoming a landscaper. Working at the greenhouse was a real bonus as I got my pick of plants and trees that came in right off the truck. After they were deducted from my meager paycheck, I hardly made lunch money! I also got to adopt and rescue (for free!) plants that were becoming rootbound in the pot or whose health was becoming endangered by baking all day in too-sunny displays. I loved saving their lives and felt I was genuinely contributing to the health and beauty of the planet.

Show me your garden, provided it be your own, and I will tell you what you are like.
ALFRED AUSTIN

After a long summer of working, swimming, and gardening at home, I was glad to get back to the fall routine of school. I loved watching the season change toward the restful cold-weather time of winter. Aside from some mulching and pruning, there was nothing for me to do but sit back and watch the changes in my family's yard. My family was somewhat stunned by my gardening industry. With my supplies from the greenhouse, I had transformed an "okay" lawn into an impressive showplace, complete with a little Japanese multilevel Zen garden with a Japanese maple, a pond that didn't hold water very well, and the pièce de résistance—a chipped pagoda I'd gotten at the nursery!

When the winter snows hit, most of my efforts were hidden from view as my treasured plants took a well-deserved rest. After a two day blizzard, everything was under a white, fluffy blanket. My little Zen garden, however, still drew the eye. Finally, I understood the wisdom of Oriental gardeners. Their gardens were created to have changing beauty throughout all the seasons. I had only imitated what I had seen in the gardening books at the nursery. It was only in the dead of winter that I realized the red bark and leaves of the Japanese maple were stunningly beautiful against the blanket of snow. The Pyracanthus climber I had planted was such a dark green as to almost be black with blazing orange berries while the evergreen shrubs, growing free form, took on an entirely eastern aspect next to the pagoda. Just looking at my little Zen garden in the snow filled me with inner peace.

ZEN CENTERPIECE

Truly nothing could be easier than this arrangement; it will foster serenity wherever you place it.

Small dark rocks
Colander
Shallow bowl
3 small floating candles
1 flower such as a gardenia, rose, or hibiscus

Place the rocks in a colander and rinse. Take the bowl and fill the bottom with 1–2 inches of rocks, depending on the depth of the container—you want to create a rock bottom. Fill with water up to 1 inch from the top. Float the candles and gently place the blossom on the water and allow it to float.

THE GARDEN GROWS YOU

Several years ago, I was walking in March along a gravel road that led to the ocean in Rhode Island. A very old and very thin woman came hobbling down a driveway toward me. I waved and continued walking, but as I passed, she grabbed my arm, turned around and began to pull me in the direction of her house. I instantly thought of the witch in *Hansel and Gretel,* and tried to pull back, but that only made her clutch tighter around my wrist. Besides, she didn't cackle, so I relented.

What was paradise but a Garden?

WILLIAM COLES

She didn't say a word, in fact, until we approached her house: a shingle-style cottage with green shutters and a front lawn erupting everywhere in purple crocus. She released me there, throwing her arms up in the air and shouting, "Look at this splendor! Isn't it a miracle?!"

I didn't know what to say. I mean the crocus were everywhere and indeed they were beautiful, but a miracle? Besides I had a very important problem I had been thinking about when she interrupted me and . . .

Betts threw her arms around me. She smelled like earth and onions, rain and soap. She whispered in my ear, "You don't understand. As you grow a garden, it grows you!"

After that, I visited Betts Wodehouse every chance I could. It turned out she was a famous sculptress. Her mother had been friends with Rodin. She had letters from him in her study, but I never went inside Betts' house. We always went directly to her garden. At 90 or so, it was hard for her to bend down and weed, so I'd do it for her. One day she talked on and on about this weed that had spread everywhere and couldn't be pulled out without also taking some of the healthy plants. I realized later, she was also telling me she had incurable cancer.

Each time we were together, she taught me about gardening, but when I arrived home, I knew she had also been offering me lessons of the soul, lessons to nourish my dream seeds, especially the twisted ones. As I would walk away down that gravel driveway, she'd always call to me, "Don't forget, as you grow your garden, your garden grows you!"

PRESSED FLOWERS

Making pressed flowers is incredibly easy. It requires no special equipment and costs absolutely nothing. Here's how: When your new telephone book comes, save the old one and put it somewhere where you won't lose it. Find a meadow and collect small bouquets of wild flowers. Lay them flat in different parts of the phone book. Place a small boulder, or anything else that's heavy and not likely to take off, on top of the phone book. Let sit for a few months.

PLANTING THE SEED

If there is anything more therapeutic and satisfying than working in a garden, it's planting the seed of wonder in the mind of a child.

And all it lends to the sky is this—
A sunbeam giving the air a kiss.
HARRY KEMP,
"THE HUMMINGBIRD"

Remember playing with puppetlike snapdragons when you were a kid? It's not a lost art. Youngsters may also enjoy fast-growing, towering sunflowers, fuzzy lamb's ears, fragrant sweet peas, lavender, and mint. And with a strawberry in a pot in a sunny location, kids can have their gardening project and eat it, too.

Or help youngsters sprout seeds indoors. Fold a couple of paper towels together to form a strip as wide as the towel and a few inches high. Moisten and place inside a peanut butter jar or similar size jar, forming a border at the base. Crumple and moisten another paper towel and stuff into the center. Carefully place seeds—beans are easy to grow and handle—between the folded paper and the glass. Keep moist, but not soaked—for several days as seeds germinate. Kids can watch roots grow and plants sprout. When plants reach above the jar and two sets of leaves have formed, transplant to pots of soil or into the ground.

MEANINGFUL STONES

Instead of buying gravel or garden rocks, consider collecting what you need from beaches, mountains, river beds, etc. Each hand-selected rock will mean more—and be more beautiful—than a pile of uniform stones.

THE WONDERS OF WILDFLOWERS

My mother is a naturalist at heart. She treasures wildflowers much more than the domesticated plants I adopted as a child. She would take me on wildflower walks and teach me the secret flora of meadow and wood. I learned to identify wild irises, jack-in-the-pulpit, Dutchman's breeches, larkspur, lady's slippers, and dozens of gorgeous and delicate specimens. I wondered at the difference between the small and seemingly rare wildflowers and the big and bold flowers that grew in our garden. The irises were in great contrast—wild irises were about four inches high and the irises I started from my aunt's were over two feet tall.

They tell us that plants are perishable, soulless creatures, that only man is immortal, but this, I think, is something that we know very nearly nothing about.
JOHN MUIR

One day, I decided to surprise my mother by transplanting some of her treasured wild irises to a flower bed at home. She was pleased, but warned me that these delicate plants simply wouldn't thrive outside their habitat. By the next spring, however, we had hearty clump of wild irises growing beside the shameless "flags" from Auntie's house.

GREETING CARDS

Place pressed flowers in a pattern you like on the front of blank cards or on stiff artists' paper you can get at a craft or variety store. Attach them to the paper with a dab of glue. Peel an appropriate amount of transparent, self-stick plastic film (like contact paper) from the roll and carefully place on top of the flowers, pressing from the center to the edge to eliminate air bubbles. Trim the edge of the plastic to match the card or paper. You can then send them to your friends for Christmas, birthdays, Valentine's Day, or no reason at all. Bookmarks can be made in exactly the same way—just cut the paper to an appropriate size.

THE DELIGHTS OF FANTASY

To see a hillside white with dogwood bloom is to know a particular ecstasy of beauty, but to walk the gray Winter woods and find the buds which will resurrect that beauty in another May is to partake of continuity.
HAL BORLAND

Winter can be hard on gardening fanatics, forced indoors to attend only to the houseplants. So I always cheer myself up by collecting all the bulb and seed catalogs that come throughout the year and saving them for a dreary January weekend. I sit down at the kitchen table with them all spread out in front of me. First I pore over the beautiful color pictures and the accompanying descriptions, fantasizing about the incredible garden I could have if money, time, and weather conditions were no object: Shirley tulips "ivory white with purply pictee edge," a one-of-a-kind Batik Iris that has "dramatic white spatters and streaks against a royal purple ground," *Alboplenum,* "doubly rare for being both multi-petaled and white."

After I have completely satisfied my eyes, I get real. I make a map of my vegetable and flower gardens, check out what seeds I have left from last year, and plot out what I want to plant. Then I go back over the catalogs again with a more selective eye and choose what I really need. Often this process goes over many days and both parts give me great pleasure: the indulging of my wildest gardening fantasies, and the anticipation of color, beauty, and form in my actual garden.

A GARDEN JOURNAL

You don't have to limit yourself to just the facts in your garden journal—it can also be a place to muse, collect quotes, and keep in touch with nature's wisdom. The key is to recognize that it can be a visual record as well as a written one—dry and paste the first sweet pea your son grew; the photos of your orchid cactus in bloom; a smattering of fall leaves on the day you found out you were pregnant surrounding the poem your husband wrote on the occasion; sketch the color of the sky on a memorable winter day.

Workshop leader Barry Hopkins, who calls these Earthbound Journals, suggests you start with a blank, hardcover artist sketchbook, at least 7½ x 8½, to allow room to paste things in, a small utility knife (such as an X-acto) for cutting, watercolors for borders, oil pastels for flowers, etc., aerosol glue for pasting, and fixative to keep pencils and pastels from smudging.

It's important that you make your own cover for the Garden Journal—you can use old bits of a special shirt, for example. If you have an electric drill, you can drill through the book to create a ribbon or rawhide fastener. The point is to follow your own creativity to where it leads you and to imbue the book with the memories of the garden.

LEI DAY

In Hawaii, a lei is given on important occasions—weddings, birthdays, housewarmings. Sometimes I choose the flower of the Hawaiian *alli* (royalty), the beautiful and delicate *ilima,* millions of petals of which are required to string a single lei; other times the richly fragrant leaves of the *maile* vine, the wearing of which traditionally symbolizes victory. The sultry scent of the fragile *pikake* (Tahitian ginger) and the sweet, honeyed aroma of the common *plumeria* (frangipani) combine with the fragrances of countless other flowers, creating a sultry perfume that is the essence of Hawaii.

Everything in nature invites us constantly to be what we are.
GRETEL EHRLICH

HOMEMADE LEIS

Leis can be easily strung using a variety of flowers. Some of the best are: tuberoses, roses (use buds, not open flowers), carnations, and asters. You will need a long (1 to 2 inches) sewing needle, and unscented dental floss, and at least three dozen blossoms of any of the above flowers. Measure a length of the floss around your neck; leis are most comfortable when they reach nearly to your natural waist. String a bead or make a large knot in one end of the floss. Trim each flower so that only the blossom remains; then, starting from the stem end, insert the needle and push through the center of the blossom, and out the other end. (You may be more comfortable wearing a thimble.) Continue until the string is full, with $1/4$ inch remaining to be tied off (you will probably need more flowers than you think!). Leis can also be made with paper flowers or candy; even dry cereal! They make excellent centerpieces, too.

FOUR O'CLOCKS

Some flowers are pure magic. I first learned that when I got a packet of flower seeds for a 4-H project one spring. The flowers were called "Four o'Clocks" and, as evidenced by the brightly colored picture on the front, were quite showy. I eagerly dug up a bed in front of my house and sowed the seeds in wobbly rows. Every day, I ran out to check on the progress, which, of course, because I was an impatient eight-year-old, wasn't quick enough for me.

Finally, after an agonizing couple of weeks, the seedlings came poking through. They grew pretty rapidly—those that could survive my overwatering. I must confess that I had gotten bored with the plain green seedlings; then the first flower buds appeared. They all burst into bloom on practically the same day, filling the front of my house with a riot of color. I, too, was bursting with pride and made all my family and neighbors look at the amazing miracle the seed packet had produced. By the time I had rounded everybody up, however, the flowers were closed up tightly. Every one of them! "That's why they're called four o'clocks," my mother explained. "Every day they close up at Four o'Clock sharp and open up with the first rays of sun in the morning."

I checked every day and she was right. Four o'clock sharp. You could set your watch by my flowers!

Never give up listening to the sound of birds.
JOHN JAMES AUDUBON

A BLOOMING CARD

Someone figured out how to plant seeds inside paper and now there are a variety of places you can get wonderful cards and notepaper that can be written on, mailed, and then planted either inside or out by your lucky letter recipient. Some of the best ones I've seen are done by the Santa Fe Farmer's Market Cooperative Store and are handmade from organic and recycled materials and studded with flowers, herbs, and vegetable seeds. The paper disintegrates when you plant it, and the seeds blossom into evidence of your feelings. Available from Seeds of Change (see Resources section on page 124.)

THE PROBLEM OF PLENTY

No gardener worth his or her salt would be without a plethora of recipes to handle the abundance of green beans, corn, tomatoes, cucumbers, and, especially, zucchini that even a tiny garden can produce. Here are two particularly tasty ones.

STUFFED GARDEN

This is a traditional recipe in Spain. You can make it with all one vegetable, but then you'll have to call it something else.

2 large onions, peeled but left whole
2 large whole green peppers
2 medium zucchini
Pinch of saffron
2 bay leaves
1/4 teaspoon nutmeg
2 cloves
2 large fresh tomatoes
2 cloves garlic, peeled and chopped
14-ounce can chopped tomatoes

4 tablespoons olive oil
Salt and pepper
4 ounces bacon, trimmed of fat (can be
    omitted for vegetarian version)
2 tablespoons lightly toasted pine nuts
3 tablespoons bread crumbs
1 cup cooked rice
2 tablespoons white wine or water
4 tablespoons grated Parmesan cheese

Blanch the whole onions, peppers, and zucchini in a large pan of boiling water, removing peppers and zucchini after 5 minutes and onions after 15. Allow to cool.

continued

STUFFED GARDEN

continued

Pour off all but ½ cup of the water and add the saffron, bay leaves, nutmeg, and cloves and simmer for 10 minutes. Set spiced water aside.

Meanwhile, using a sharp knife and a spoon, cut the tops of the tomatoes and scoop out the insides, leaving a shell thick enough for stuffing. Chop the tomato insides. In a small saucepan, combine 1 tablespoon oil, the garlic, and all of the chopped tomato—fresh and canned. Cook over medium heat for 10 minutes, until it becomes saucelike. Add salt and pepper and set aside.

Preheat oven to 400°F. Using a spoon and a sharp knife, scoop out the center of the onions, leaving a thin outer shell. Chop the centers. Cut the zucchini in half and scoop out the center, again chopping the insides. Cut the tops off the peppers and remove seeds and membranes, keeping the peppers whole.

Heat the rest of the oil and the bacon, sauté chopped onions, and zucchini until onion is wilted and bacon is crispy.

In a large bowl, combine the bacon mixture with the pine nuts, 2 tablespoons of bread crumbs, half of the tomato sauce, salt and pepper, rice, and wine or water. Place the vegetable shells in a large baking pan and fill with the stuffing mixture. Top with remaining tomato sauce, bread crumbs, and grated cheese. Carefully pour the spiced water into the pan so that vegetables are sitting in about ½ inch of water. Bake until vegetable shells are soft to touch and stuffing is heated through, about 35 minutes. Add more spiced water to keep from scorching if necessary. Serves 8 as a side dish; 4 as a main course.

*Tickle it with a hoe and
it will laugh into a harvest.*
ENGLISH PROVERB

ROASTED TOMATO AND RED PEPPER SOUP

2 1/4 pounds tomatoes, halved lengthwise
2 large red bell peppers, seeded and quartered
1 onion, cut into thick slices
4 large garlic cloves, peeled
2 tablespoons olive oil
Salt and pepper
1 teaspoon fresh thyme leaves, or 1/2 teaspoon dried
Water

Preheat oven to 450°F. Arrange tomatoes (cut side up), bell peppers, onion, and garlic cloves on a large baking sheet. Drizzle oil over all; sprinkle generously with salt and pepper. Roast vegetables until brown and tender, turning peppers and onion occasionally, about 40 minutes. Remove from oven. Cool.

Transfer vegetables and any accumulated juices to food processor and add thyme. Purée soup, gradually adding about 2 cups of water to thin soup to desired consistency. Chill until cold, about 3 hours. (Can be prepared 1 day ahead. Cover and keep refrigerated. If soup becomes too thick, thin with water to desired consistency). Serves 4.

THE BOLDNESS OF TULIPS

I love tulips better than any other spring flower. They are the embodiment of alert cheerfulness and tidy grace, and next to a hyacinth they look like the wholesome, freshly scrubbed young girls beside stout ladies whose every movement weighs down the air with patchouli. Their faint, delicate scent is refinement itself; and is there anything in the world more charming than the sprightly way they hold up their little faces to the sun? I have heard them called bold and flaunting. But to me they seem modest grace itself, only always on the alert to enjoy life as much as they can and not afraid of looking the sun or anything else above them in the face.

For all things produced in a garden, whether of salads or fruits, a poor man will eat better that has one of his own, than a rich man that has none.
J. C. LOUDON

THE BULB GATHERER

In the fall, I go out in the garden and dig in the dirt for treasure: clumps of bulbs that were only a single bulb in the spring. I divide them up and find a new home for each bulb to start another family. Then I do my best to forget where I've put them so I can be pleasantly surprised in the spring. If everyone did this conscientiously, the whole earth would eventually be covered in daffodils and tulips and hyacinths. Squirrels can't do it all by themselves. What the world needs now is a Johnny Tulip bulb.

SUMMER'S HEALTHFUL HARVEST

We know gardening is good exercise, but research is now showing that eating the bounty from our gardens is also good for our health. Here's the lowdown on the most popular summer crops:

Tomatoes: One medium tomato has half the (RDA) recommended daily allowance of vitamin C and 20 percent of the RDA of fiber and vitamin A. It also contains lycopene, an antioxidant that appears to reduce the risk of heart attack and of various cancers, including breast and prostate.

When the world wearies, and society ceases to satisfy, there is always the garden.

MINNIE AUMONIER

Corn: Excellent source of fiber and two antioxidants—lutein and zeaxanthin—that may lower the risk for macular degeneration, which is the leading cause of blindness in older folks.

Sweet Peppers: These are little powerhouses of vitamins A and C. Green peppers have twice as much vitamin C as oranges; yellow and red peppers have four times the C! They also are good sources of immune system enhancing B-6 and folic acid, which helps against heart disease and has been recently discovered to prevent neural tube defects in developing fetuses.

Zucchini: You may get sick of eating it, but this low-cal wonder is full of vitamins C and A and is also high in fiber, which helps prevent heart attacks and colon cancer.

The work of a garden bears visible fruits—in a world where most of our labors seem suspiciously meaningless.

GREEN BEAN, CORN, AND TOMATO SALAD

1 pound green beans, cut into 1-inch lengths
Kernels from 3 ears of corn
1/2 cup white wine vinegar
6 tablespoons olive oil
5 tablespoons sugar
3 large tomatoes, chopped
1/2 cup chopped red onion
1/3 cup chopped fresh parsley
Salt and pepper

Cook green beans in a large saucepan of boiling water for 2 minutes. Add corn kernels and cook until vegetables are crisp-tender, about 2 minutes longer. Drain well.

Whisk vinegar, oil, and sugar in large bowl to blend. Add beans, corn, tomatoes, onion, and parsley; toss to coat. Season with salt and pepper. Cover and chill at least 2 hours or overnight. Serves 6.

THE PLEASURES OF PERIWINKLE

Growing older has shown me the value of ground covers: hale survivors they are. Ground covers do just what their name promises, at varying rates of speed. My favorite is periwinkle or *Vinca minor*—deep evergreen foliage that vines and hugs the ground and will grow up a hill or along a wall in the poorest of soil. I placed one tiny plant

The sound of birds stops the noise in my mind.
CARLY SIMON

alongside the chimney of my house, a place that had defied other plants. It thrives there still and has spread over the years to cover an entire hill. A few years ago, I undertook a scholarly research project about *Gawain and the Green Knight,* the epic poem in Middle English. To my great surprise and delight, I discovered that mysterious women had embroidered periwinkle onto Gawain's cape before his heroic journey to confront the monstrous Green Knight. As I dug deeper in the arcana section of my library, I discovered that periwinkle had been made into wreaths and placed on the heads of young men going to battle in medieval England. What a rich history my favorite humble ground cover comes from!

A TIARA OF FIREFLIES

Our grandma always wore hairnets to keep her silver hair in place. As the nets wore out or snagged, Gram would give them to us girls. On special evenings my sis and I would weave garlands of daisies and wildflowers and pin them in our hair. Our brother would bring in a canning jar full of fireflies, and we would tip it into Gram's hairnets, loosely fit them over our fancy hairdos, and have twinkling tiaras for an evening of garden play.

GHOST GARDEN

Where my mother's family comes from, there are many ramshackle houses. In shades of ghostly grey, their inhabitants have grown up and moved out, leaving the sad, spectral houses behind. The luckier of these houses, along the banks of the Ohio River, get reclaimed and become showplaces, at least one of which has graced the pages of *Architectural Digest*. Mostly though, they wait. While the houses sag and creak, the gardens go wild! Plantings and hedges, once neat and tidy, are out of control, buzzing with birds, bees, and heaven knows what. Every spring and summer as a child, I would pick flowers in these abandoned yards and set up a table at the bottom of my driveway to sell bouquets in old coffee cans to passing neighbors.

Patience is a flower that grows not in everyone's garden.
ENGLISH PROVERB

One such house I know burned to the ground during the night, leaving a charred and barren lot where once had been some river captain's pride. I was sad because the hydrangeas were gone, and I had my eye on them, both for bouquets and for transplanting in my mother's garden. The next year, an amazing thing happened around the outer perimeters of the burned-down old manse: A "ghost garden" was growing in exactly the same formation! Bulbs, hedges, and even the hydrangeas all came back at an astonishing pace. And thanks to all the rusting copper pipes, hinges, and nails scattered half-buried in the soil after the fire, the hydrangeas, a pristine white the year before, were the bluest of blues. I decided not to take any. Those hydrangeas had been so loyal to that yard, they simply could not move anywhere else.

HEAVENLY HYDRANGEAS

If you have some hydrangeas of your own, late summer, when the blossoms have just started to dry on the bush but before the color has faded, is the time to harvest for drying.

Getting perfect, colorful, full, fluffy dried heads isn't completely easy, however. Flower experts report only about a 50 percent success rate, so it's best to start with a lot of flowers. Lace-cap hydrangeas are particularly difficult; try the mop-head varieties instead. The trick is to remove all leaves and recut the stems as soon as you get into the house. Hold the stems over a flame for 20 seconds, then submerge in cold water up to the flower head for 2 days. Pour off all but ½ inch of water and let the rest evaporate in a warm dry place, such as next to your water heater. When the water is gone, the blooms should be dry.

AROMATHERAPY BASICS

Aromatherapy, the use of scents from the essential oils of plants to alter mood and promote healing, is an ancient art currently enjoying a booming revival. While many common garden plants are used in essential oils—peppermint, basil, and lavender, to name just a few—the quantities of flowers or leaves needed to produce the oil (1,000 pounds of jasmine flowers for one pound of oil, for example) means that even the most prolific gardeners would be better off buying their essential oils from catalogs or stores.

To be overcome by the fragrance of flowers is a delectable form of defeat.
BEVERLEY NICHOLS

Most commonly the oils are used in the bath (put in at the very end; the water should be no more than 100°F) or in a diffuser or placed on a handkerchief and inhaled when you need a lift. Since essential oils are very potent, they should always be diluted with a base oil such as sweet almond or grapeseed before being put on your skin. And don't ingest or get it in your eyes.

If you are pregnant or have a chronic illness of any kind, consult your physician before using any.

ESSENTIAL OILS

Here are some of the most common essential oils and their qualities:

Basil: uplifting, clarifies thought processes
Bergamot: uplifting, yet calming
Cedarwood: relaxing, stress reducing
Chamomile: soothing and calming, excellent to use after an argument
Eucalyptus: invigorating, cleansing, tonifying
Fennel: relaxing, warming, calming
Fir needle: refreshing, cleansing
Frankincense: calming, helps release fear
Geranium: balancing mood swings, harmonizing
Juniper: purifying, stimulating
Lavender: calming, soothing, relaxing
Lemon: uplifting, refreshing, mental alertness
Lemongrass: stimulating, cleansing, tonifying
Lime: invigorating, refreshing
Mandarin orange: uplifting, refreshing
Marjoram: very relaxing, anxiety reducing
Myrrh: strengthening, inspiring
Orange: uplifting, refreshing
Patchouli: inspiring, sensuous
Peppermint: stimulating, cleansing, refreshing, invigorating
Pine: refreshing, cleansing, stimulating
Rose: emotionally soothing
Rosemary: stimulating, cleansing, good for studying, invigorating
Sage: cleansing, purifying
Sandalwood: stress reducing, sensuous, soothing, helps release fear
Spearmint: refreshing, stimulating
Ylang-ylang: uplifting, sensuous

PUBLIC GARDENING

Longing for a garden but have no place for one? Take advantage of the variety of places that have gardens: zoos, public gardens and parks, cemeteries, college campuses, garden club tours,

The only conclusion
I have ever reached is that
I love all trees,
but I am in love with pines.
ALDO LEOPOLD

nurseries and garden centers, or a friend's house. In many cities these days, there are also community gardens and gardening co-ops in which you can get your hands dirty. Call your parks and recreation department. (All of the above are also great places to get ideas if you do have a garden).

PRODUCE FOR APARTMENT DWELLERS

If you have limited space or time for a garden (or are plagued by critters eating your goodies before you get to them), try creating hanging vegetable baskets. According to experts, almost anything can be grown in a basket, but be sure to get compact growing varieties of the vegetables you want. Buy 14-inch diameter wire baskets (16-inch for zucchini or watermelons). It's best to grow one type of vegetable per basket, although a variety of lettuces or herbs will work well together.

Line the basket with sphagnum moss and fill with potting soil. Plant seedlings rather than seeds, and hang the baskets outdoors from patios or rafters where they will get at least four hours of afternoon sun. Avoid overwatering seedlings, but once they become established, be aware you need to feed and water frequently; on the hottest days, they may even need to be watered twice a day! Once seedlings are three weeks old, fertilize every three weeks with an all-purpose soluble fertilizer, but never feed unless the soil is damp.

SPADE WORK

I am not a patient person by a long shot. I move quickly through my day and my life, wiggle and sigh and roll my eyes when forced to stand in line, and expect instant change in those I love (including myself). Being exposed to Buddhist meditation has helped me to slow down or, more accurately, recognize the need to slow down and respect the process, not the destination. So recently I've been thinking a lot about dirt.

Dirt as in soil and the sound preparation a garden needs. In the spring it's all about spade work, studying what nutrients are needed now so that my plants will get the nourishment to really flourish. Last year, an experienced gardening friend, when confronted in the heat of summer with my drooping tomatoes, told me I needed to double dig the garden patch (go two pitch-forks deep into the soil and add compost) because the clay in my dirt was choking my plants.

Good gardening is very simple, really. You just have to learn to think like a plant.
BARBARA DAMROSCH

This spring, as I dig deeper and deeper—hardened clay is not easy to break up—I'm struck by how much work that never shows overtly goes into a successful garden. It will be months before I find out if my digging is worth it. Today I'm just digging not looking for a quick fix, adding what is needed. Ah, yes.

DETOXIFY

You don't need a mass of chemical pesticides, weed killers, and fertilizers to have a lush garden. There are a number of things you can do to reduce the amount of chemicals on your fruit, vegetables, and flowers, which in turn will reduce the amount of such items in the water and air and limit exposure to birds, bees, butterflies, your children, and yourself. First, start a compost pile; it should eliminate your need for nonorganic fertilizer. Ask your nursery about organic ways of fertilizing lawns; there are several. Learn about biological ways of controlling pests: praying mantises, spiders, and ladybugs reduce harmful insects. My only method of aphid control for two dozen rose bushes is to buy a batch of ladybugs at the beginning of the flowering season; it works like a charm. Ladybugs are available at most nurseries and through catalogs.

SUSTAINING DELIGHTS

My favorite garden memories are from the vegetable gardens my grandfather grew during the depression when I was a little girl. My grandparents had lost all their savings when the banks folded.

We spent all summer with my grandparents. My grandfather planted
a garden large enough to feed all six of us and supply my mother and
grandmother with produce to can for the winter. The garden was
beautifully kept with the lowest vegetables in the front—car-
rots, radishes, beets, parsnips. Then three kinds of beans—
green and shell (too cold for limas in their climate). The
polebeans, corn, and popcorn. Off to the side was a ram-
bling mass of cucumbers and squash. Pumpkins were planted
in between the last two rows of corn.

*Sow the living
part of yourselves
in the furrow of life.*
MIGUEL DE
UNAMINO

Everyday the noon meal consisted of any and all vegetables that
were ready. And I always rode home at the end of the summer in the
backseat of the car squeezed between boxes of full canning jars.

DETOXIFY

There are all kinds of bug deterring herbs that can be planted as companions to various
vegetables and flowers: Marigolds are the workhorse here; they deter most bugs. Garlic
should be planted between rose bushes to keep away Japanese beetles. Basil repels flies
and mosquitoes and helps tomatoes grow. Horseradish helps keep potato bugs away.
Mint and peppermint deter white cabbage moths. Ask your nursery for a complete list.

THE SHAPE OF THINGS

At times, it seems to me quite an obvious sort of pleasure. Yet I still find the shapes and colors of flowers to be truly amazing. Last year I decided to plant some flowers to attract butterflies and hummingbirds. One was a cleome, which I had never tried before. Unfortunately I planted it in the early fall, just after the heat of summer, hoping to have flowers fairly quickly. (I am still new to the art of gardening and have not learned to trust that other people know a lot more than I do. This is a double-edged sword. On the one hand, I tend to begin things in the wrong season with little success even though someone has told me it won't work. On the other, I find I can grow some wonderful plants that catalogs and books say are not possible in my area. So, I keep experimenting.)

It is the leisure hours, happily used, that have opened up a new world to many a person.
GEORGE M. ADAMS

Well, four cleomes sprouted and began to grow. I was very optimistic and was looking forward to their flowering by Thanksgiving. I waited and waited and waited. The plants grew to about a foot tall by Christmas and then decided to lay low for the rest of the winter. Discouraged, I wondered if any bloom would ever show up. After the last frost, I transplanted them into a space in the garden and waited again. Finally they grew a bit more but it wasn't until June that the growth really took off. Then within a month I had buds and my first flowers. What incredible flowers. Solid, pure tones of pink or white with a shape that seems to be from science fiction: Four oval petals in a vertical fan shape with the pistil and five or six stamen coming way out from the center of the flower like Martian antennae. Truly an out-of-this world beauty, worth all the wait.

WEEDING VACATION

It was April and I was just plain exhausted from the effort of keeping my business afloat. I couldn't afford to go anywhere, so I decided to take the week off and do nothing. I slept, read, watched videos. But after a few days, I got a bit bored and so my husband and I decided to tackle the front yard, which had been ignored since we moved the previous summer. The dry creek made of stones that dissected the property was overgrown with strawberrylike plants that yielded no edible fruit and seemed to smother everything else in sight. The previous owners knew nothing about plants and they had put shade-loving ones in full sun and sun-lovers in deep shade. Things were too close together; an ugly boxwood hedge rimmed the entire perimeter. . . . It needed work.

A violet, by a mossy stone
Half hidden from the eye!
Fair as a star, when only one
Is shining in the sky.
WILLIAM WORDSWORTH

My husband tackled the bigger project—moving various plants around—and I set to weeding the dry creek. Maybe it was because, like most of us in the modern world, I work in the intangible world of ideas all year, but as I sat there weeding patch after patch, exposing the stones as I worked, I felt an inordinate sense of satisfaction. It was so concrete: you pulled a weed, and then another and pretty soon a whole area was pristine and perfect. If only the world of business could be as tidy. Granted, later you'd have to do it all again, but at least you could see immediately the effects of your labor in stark brown earth. I enjoyed it so much that now I purposely take weeding vacations.

SALT GLOW

This is fabulous for exfoliating dead skin, particularly when your tan is starting to flake. Be sure not to use on your face or neck; it's too rough for that.

2 cups sea salt
7 drops of your favorite essential oil
1 ounce sweet almond oil

Place salt and oils in a bowl and combine well with your fingers. Stand or sit naked in an empty bathtub and rub salt mixture into your skin with your hands, starting with your feet. Massage in a circular motion. As the salts fall, pick up and reuse until you reach your neck. Then fill the tub with warm water and soak.

CRISP DELIGHT

*The best things
that can come out
of the garden are gifts
for other people.*
JAMIE JOBB

One of my greatest garden pleasures is to receive compliments on the rhubarb-strawberry crisp I bake for almost every summer barbecue. Our rhubarb plant was well established in the garden when we moved into our house fifteen years ago. The plant is so prolific that I give away a fresh bunch to every rhubarb lover who enters the door. Though not all our guests favor this beautiful celerylike plant topped with magnificent poisonous leaves, when they see the crisp warm from the oven with its sugary top and juicy bottom, they cannot resist this old standby. Add ice cream and wait for the compliments to come pouring in.

SUREFIRE RHUBARB STRAWBERRY CRISP

3 cups rhubarb, sliced
2 cups strawberries, whole or sliced
Juice of 1 lemon

1 stick butter, softened
1 cup granulated sugar
1 cup flour

Preheat oven to 400°F. Combine rhubarb, strawberries, and lemon juice in a 9 x 13-inch baking pan. In a medium bowl, combine the butter, sugar, and flour until crumbly and then spread over rhubarb mixture. Bake uncovered for 20 minutes or until the crisp is bubbly and the top is browned. Serves 6.

THE SATISFACTIONS OF HERBS

On a warm summer's day, the delightful scents of rosemary, mint, and lavender perfume the air as bees and hummingbirds go about their feeding. Perhaps you harvest a handful of lavender to scent your underwear drawer. Later you run out to snip some parsley and chives to add zest to a salad and throw together a bouquet of bright-red bergamot and silvery artemisia for the dinner table. Or maybe today is the day you create fragrant potpourri and dried wreaths from the bounty of your garden to please the eyes and noses of friends and family.

Nothing is more pleasing, useful, and easy than growing an herb garden, even if you only have a tiny plot of land. You don't even really need a yard, because herbs can be easily grown in pots on a deck or terrace or even indoors if you have a sunny window.

Innumerable as the stars of night, Or stars of morning, dewdrops which the sun Impearls on every leaf and every flower.
MILTON

If you do have outdoor space, tuck them in the border of a flower garden or in a patch in the vegetable garden or simply plant them directly outside the kitchen door, like they do in France, so you just lean outside when you want something. Because they are virtually indestructible, they are perfect for beginning gardeners. Most are fast growing and pest resistant. Chances are your only problem may be how quickly they spread; some can be quite invasive.

Choosing what to grow is a pleasure in and of itself. Think about what and how you cook and whether you would like to make your own flower crafts and potpourris. Then pick accordingly. I personally hate to pay for a bunch of dill at the store because when I need it, I want only a little, and the rest always spoils in the fridge. So dill was at the top of my list. Some other common culinary herbs to consider: basil, chives, marjoram, mint, oregano, parsley, rosemary, sage, savory, and thyme. If you want a scented garden and plan to make bouquets, potpourris, wreaths, etc., you might want to plant bergamot (beautiful shaggy red flowers), dyer's chamomile, lady's mantle, lamb's ears, lavender, lemon balm, lemon verbena, nasturtiums, rue, santolina, tansy, violets, woodruff, and yarrow.

ROSEMARY RINGS

For a holiday party, decorate your serving platters with rosemary rings. Simply shape long branches into a wreath and tie with floral wire. Place on the plate and decorate branches with cherry tomatoes. Put food in the center. You can also use shorter rosemary branches to make festive napkin rings; again secure with floral wire.

REMEMBERING LILACS

Unless the soul goes out to meet what we see we do not see it; nothing do we see, not a beetle, not a blade of grass.
WILLIAM HENRY HUDSON

I suppose the garden behind my grandparents' house was small, but to a four year old it seemed immense. The distance from the backdoor to the end of the yard was a journey from the safety of home, across an expanse of grass, around orderly flower beds, and finally to the marvelous wilderness of the tall, old lilac hedge. I discovered that a persistent push would let me enter a cool, green space under the branches of the lilacs. There I daily established my first household, presiding over tea parties for an odd assortment of stuffed animals and the patient family cat.

Now, nearly seven decades later, the heady scent of lilacs takes me back to that garden where I took those first ventures toward independence—though never out of sight of the familiar backdoor.

BASIL, MINT, AND ROSE HIP TEA

12 ounces water
2 tablespoons chopped fresh spearmint
2 tablespoons chopped fresh basil

2 rose hip tea bags
Honey or sugar

Place the water, spearmint, and basil in a nonreactive pan. Cover and bring to a rolling boil. Remove lid, stir, then add the tea bags. Steep, covered, for 3 minutes. Strain tea into 2 warmed cups. Sweeten with honey or sugar, if desired. Serves 2.

EDIBLE BLOSSOMS

Yes, many flowers (minus stems and leaves) are quite wonderful tasting. But before you start randomly eating flowers from your garden, be sure you know what you are doing—some are deadly poisonous. And of course, if you use pesticides or herbicides in your garden, do not eat unwashed blooms. Caveats aside, flowers do wonderfully in salads, as a garnish to a cold summer soup, as garnishes for serving platters, and to decorate cakes.

The following is a list of some of the edible beauties:

bee balm

calendula

Everybody needs   daylilies

beauty as well as bread,   hollyhocks

places to play in and pray   marigolds

in, where Nature may heal   nasturtiums

and cheer and give strength   pansies

to body and soul alike.   roses

JOHN MUIR   scarlet runner bean

sunflowers

violets

CANDIED FLOWERS

These delectable treats are easy to create; use them on top of ice cream or cakes. Pick the flowers fresh in the early morning.

A generous handful of violet blossoms, rose petals, or any flower from
    the edible flowers list
1 or 2 egg whites, depending on how many flowers you use
Superfine sugar

Gently wash flowers and pat dry with a clean towel. Beat the egg whites in a small bowl. Pour the sugar into another bowl. Carefully dip the flowers into the egg whites, then roll in sugar, being sure to cover all sides. Set flowers on a cookie sheet and allow to dry in a warm place. Store in a flat container with waxed paper between layers. These will last for several days.

MY CUTTING GARDEN

One of the little things I take true delight in on a regular basis is flower arranging. I'm not good with my hands (my sister used to leave the house in fear when I was learning to sew in junior high), but I have the urge to make something beautiful, and over the years I have discovered that flower arranging is my creative medium. It takes no manual dexterity and, since I planted a cutting garden in my backyard (I can't stand the prices at the florist), no money either. I have an entire cabinet filled with containers (vases I have been given over the years, the beautiful blue bottle a certain type of mineral water comes in, an old jam jar) as well as things scavenged off store-bought bouquets (curly willow, pieces of floral foam, rocks I use to stabilize large arrangements). Every once in a while, I consider actually investing in this passion of mine) buying some floral frogs, getting some vases of certain heights and widths (why, oh why, are the containers I have always too tall or too short or too wide?). But there is something about my low-tech, New England frugal approach that I enjoy, so I continue to make do.

A garden is a private world or it is nothing.
ELEANOR PERÉNYI

I take great delight in the fact that any time of the year (ok, so I now live in California), I can walk outside and find something blooming to liven up my bedroom, the kitchen table, or the shelf in the bathroom. I have never read a book about the principles of flower arranging, and I don't spend too much time on it—maybe five minutes at the most. For me, the joy comes from the ease with which it is possible to make something pleasing to look at: selecting an old yellow mustard jar, filling it with nasturtiums, and placing it on the kitchen table. Ongoing beauty, meal after meal, in only two minutes!

LONG-LASTING BOUQUETS

When cutting flowers from the garden, there are a few tricks to gathering them that will ensure long life: (1) be sure to cut them in the early morning or evening. (2) whether using scissors or a sharp knife (there's disagreement over which is better), cut the stems at a deep angle when the buds are half open (except for zinnias, marigolds, asters, and dahlias, which should be picked in full bloom); (3) remove leaves at the bottom of the stems (they rot) and plunge flowers immediately up to their necks in tepid water; (4) let sit in a cool place for 4 to 12 hours before arranging.

Certain flowers require different treatment. Daffodils should be kept separate for 12 hours to dry up sap that clogs stems of other flowers. When using tulips or irises, be sure to remove the white portion of the stem; only the green part can absorb water. Dip stem ends of poppies and dahlias in boiling water before placing in tepid water to prevent them from oozing a substance that can clog the stems and cause the flowers to wilt.

MORNING GLORIES ALL DAY LONG

I have realized anew the almost spiritual beauty of the common morning glory. I avoided planting these flowers anywhere about the garden, because they seed so freely that they soon become an annoyance, strangling more important plants and even tangling up the vegetables mischievously. Instead, I have given them a screen that breaks the bareness of the tool house, and let them run riot. The flowers are as exquisite in their richly colored fragility as if Aurora, in the bath, had amused herself by blowing bubbles. These, catching the sunrise glow, floated away upon the breeze and, falling on a wayside vine, opened into flowers that from their origin vanish again under the sun's caress.

The earth laughs in flowers.
RALPH WALDO EMERSON

Among all their colors none is more beautiful or unusual than the rich purple with the ruddy throat merging to white—night shadows melting into the clear of dawn.

SKIN SOOTHER BATH

This wonderful recipe will soothe any skin condition, from heat rash to chicken pox. It's wonderful for your skin and hair, so use it even when you're problem-free!

½ cup finely ground oatmeal
1 cup virgin olive oil

2 cups aloe vera gel
20 drops rosemary or lavender oil

Combine the ingredients in a large bowl; stir well. Add to warm, running bathwater.

SKIN REMEDIES

You can grow and harvest a variety of herbs that will beautify and balance your skin in the bath, as well as release a variety of aromas that have various aromatherapy effects.

Everything in nature invites us constantly to be what we are.
GRETEL EHRLICH

Simply dry the herbs during the fall, and then place in a cheesecloth or muslin bag, tie it off, and drop into a tub of warm water. If you don't want to grow your own, you can order all of these plus hundreds of other herbs, muslin bags, aromatherapy oils, and books on herbs from several catalog or Web sites (see Resources section on pages 122–125). Here's some common herbs and their suggested uses for different skin types:

Dry, sensitive skin	borage, comfrey, elder flower
Mature skin	chamomile, gingko, horsetail, lavender
Oily, blemished skin	calendula, echinacea, goldenseal, myrrh, sage, yarrow

NATURE'S GARDEN FACIAL SCRUB
for medium to dry skin

1 medium-size banana	2 egg whites
6 strawberries, hulled	2 tablespoons nonfat yogurt

Blend ingredients on high for 3–5 minutes. Apply to damp skin, and leave for 10–15 minutes.

SANDWICH MASK

½ cup mayonnaise
1 large tomato
1 medium-size cucumber, chopped
½ large avocado, mashed

Purée the ingredients in a blender or mixer until smooth. Apply to skin and allow to dry. Rinse well and follow with a light toner and moisturizer.

WHEN YOU WATER, WATER

It's the leisure hours, happily used, that have opened up a world to many a person.
GEORGE M. ADAMS

Every year the sitting area on my deck shrinks as I buy more pots and fill them with summer flowers. I water them slowly, pot by pot. Once in a while, I try to do two things at once—water and talk on the phone. But it's distracting to pay attention to the person at the other end while I'm watching the water flow into the rich brown soil and smiling at the flowers. So by the beginning of July, I just let the phone ring when I'm watering.

As each plant soaks up the wet nourishment, I stand amid the air, the smells, and the sounds of life, and all of them help to slow my world down, quiet my thoughts, and give me time to pause.

LAVENDER BATH OIL

Here's another great way to relax—an aromatherapy bath.

1 cup almond or grapeseed oil
½ teaspoon lavender essential oil
¼ teaspoon vitamin E oil

Dried lavender sprigs
10-ounce decorative bottle with a top
Ribbon and gift tag, if desired

Combine the oils in a glass container and test the scent on your skin. (You might want to add a bit more of one thing or another depending on the fragrance.) Place the lavender sprigs into the bottle. Using a funnel, pour the oil into the bottle and close the top. Store in a cool, dry place.

ZILLIONS OF ZINNIAS

Fifteen years ago, my best friend Ann and I prepared a meal for an older couple in our community. As we delivered the meal,

Who loves a garden loves a greenhouse too.
WILLIAM COWPER

we paused to admire the husband's zinnias. Despite his age and weakening health, Mr. Flynn successfully cultivated a bed of zinnias each year. And what a beautiful zinnia garden it was! A rainbow of color, the bed extended the length of his driveway.

"Let's get you some flowers to take home," he said, noting our appreciation. A fondness of flowers was one of many interests Ann and I shared. No wonder we became friends when she moved to town.

As we talked, I learned that Mr. Flynn had grown his zinnias from seed. "You can save the old flowers for next year's seed," he said, as he clipped a bouquet and handed it to me.

He led Ann and me into his garage. "Look at this," he said, picking up a big box of dried zinnia flowers. "You have to keep the old flowers cut back if you want to get lots of blooms," he explained. He got two bags, filled them with zinnia seeds, and instructed us to plant them the next spring.

By spring, Ann had moved away—ten hours away. Within several years, Mr. Flynn was gone too. Some years later, my husband and I visited Ann and her family in Mississippi. As Ann and I walked through her yard, I saw a small patch of zinnias. "Remember Mr. Flynn's zinnias?" I asked.

Ann laughed. "These are from Mr. Flynn's zinnias," she said. "Do you still have some?" I was ashamed to admit that mine had fizzled out. I had been less diligent than she in keeping seeds each year. "We'll just have to send you home with some seeds," Ann said.

A single seed goes a long way, I thought. It can carry flowers and joy from North Carolina to Mississippi and back to North Carolina again. Not only had our friendship endured Ann's move, but so had Mr. Flynn's zinnias and our memory of this gentle man. These were surely friendship flowers.

TRADING PLEASURE

Start a bulb-and-seed exchange with friends. When you are doing your harvesting in the fall—dividing bulbs and drying out seeds for next year—try trading with friends for a no-cost way to increase the variety in your garden. We started doing this years ago when we found out the hard way that a packet of zucchini seeds was far too many for two people to plant and eat. We divided them up among our friends around the country and that got the ball rolling.

To send bulbs, place them in a paper bag and then in a box. To collect seeds, shake the flower heads over an empty glass jar. To send seeds, take a small piece of paper and make a little envelope out of it by folding it in half. Take each side and fold in about $\frac{1}{2}$ inch toward the middle. Tape those two sides, place the seeds inside the opening at the top and then fold the top down and tape again. Write on the outside what is inside, and mail in a padded envelope. A sweet surprise for family or friends.

IN WINTER GARB

Sometimes I think the garden is even more beautiful in its winter garb than in its gala dress of summer. I know I have thought so more than once when every leaf and branch was clothed with a pure garment of snow, so light as not to hide the grace of form. But nothing, it seems to me, could ever transcend the exquisite beauty of the vegetation when on one occasion a sharp frost followed a very wet fog. The mist driven by the wind had imparted a coating of fresh moisture, evenly distributed, over and under every leaf and twig inside the trees and shrubs as well as outside. The light coating then froze and left every innermost twig resplendent with delicate white crystals. It was quite different from an ordinary frost or a fall of snow, beautiful as are frequently the effects of these. But the glory of the scene reached its climax when the sun came out and the thicket scintillated from the center as well as its external surface.

Nor rural sights alone,
but rural sounds,
Exhilarate the spirit, and restore
The tone of languid nature.
WILLIAM COWPER

Its dazzling splendor could not last, and the glistening and enchanted spectacle gradually melted away before the greater and more glorious life-giving presence of "God's lidless eye." This gorgeous scene, however, has always dwelt in my memory and figures as the most glowing aspect a garden can assume at any season of the year.

I prefer winter and fall, when you feel the bone structure in the landscape—the loneliness of it—the dead feeling of winter. Something waits beneath it—the whole story doesn't show.

CITRUS TONER

For sluggish winter skin try this excellent facial toner:

1 cup mint leaves
¼ cup lemon peel, finely grated
¼ cup grapefruit, finely grated
1 cup water

Add mint and citrus peels to rapidly boiling water. Continue to boil for 1–2 minutes or until peels become soft and slightly translucent. Remove from heat. Cool and strain. Store in the refrigerator or freezer; this will last 2–3 weeks.

MEMORIES

It all started when my father died and friends gave me a white rhododendron as a sympathy gift. I noticed that every year when it bloomed, memories of my father would come flooding back. Then my husband's mother died, and friends sent a crab apple tree. When that burst into bloom, we would think of her as well. We decided to turn that whole section of our yard into a memory garden: the early blooming azalea my daughter, who lives in California, sent me for Mother's Day; the scented geraniums my old friend Kay once gave me a slip of; the weeping cherry we planted in honor of my husband's father; the clematis that was my husband's holiday gift from his sister. Now, as the seasons turn and the plants bloom in its turn, our thoughts go out to the person the plant symbolizes, and we draw him or her close once again.

A garden is a place to feel the beauty of solitude.
BOB BARNES

A BIRTH TREE

You don't have to wait till someone dies or moves away to have a plant in their honor. My husband and I planted a hydrangea that was on the altar at our wedding. It moves with us from house to house as a symbol of our marriage. And many people have planted birth trees for their children. You can involve your child in helping take care of the tree and track its growth by tying a bit of yarn to the outmost tip of a branch each fall and see where the yarn ends up after the summer.

FRAGRANT FIRES

You can bundle up mini-logs of assorted dried flowers to add fragrance to fires. Simply gather up an assortment of dried flowers and herbs still on the stems. Make into bunches about eight inches long and as big around as your fist. Tie each bundle with brown jute or raffia. When your fire is smoldering, top with a flower log and enjoy the fragrance.

When the world wearies,
and society ceases to satisfy,
there is always the garden.
MINNIE AUMONIER

HEARTH DECORATION

To make a hearth decoration, tie 4 or 5 bundles in a row and cut all the stems down to the same length. Take a length of jute 3 times as long as you want the finished hanging to be. Fold the jute in half. Place one bundle in the fold. Grasp both sides of the twine and securely tie the bundle with a simple overhand knot. Move 2 inches up the jute and tie another simple knot. This becomes the bottom point for the next bundle. Again tie the bundle. Continue until you've used all the bundles. You should have 12 inches or so of jute left at top. Fold that down in half and securely tie to the topmost knot. This becomes the hanger.

Dried greens	Glue gun
Floral wire	Dried flowers
Wire cutters	3 feet of 3-inch wired ribbon

Take the dried greens and create two bunches—the more, the bigger the swag. Intermingle the stems of the two bunches to create what looks like a bow tie. Wrap this in the center several times with the floral wire until it is fairly secure. Take the glue gun and squirt glue into the center to help hold the bundle together.

Take the dried flowers, wire, and glue onto the swag one type at a time. Balance the types as you go and glue only at the center 3 inches. When you have put on as many types as you want, wrap the center with three or four turns of wire. Twist the ends together and cut off the excess. Wrap with the ribbon, hiding the wire and glue and trim to desired length. (If you want a more rustic look, use up to two dozen or so strands of raffia). Create a wire hanger in the back by wrapping floral wire, on the back side of the swag, through and around the ribbon and wire, creating a loop. Hang from wire in the back.

GARDEN IN A JAR

The very first garden I remember was a sweet potato in a jar. My mama planted it when we were living in three tiny rooms behind our dress shop. There was no yard, no room outside even for a flowerpot. But Mama filled a Mason jar with water, propped the long, skinny sweet potato up in the glass with a trio of toothpicks, and told my brother and me to watch.

The hours when the mind is absorbed by beauty are the only hours when we really live . . .
RICHARD
JEFFERIES

By summer, the kitchen window was curtained with graceful vines and big curving leaves. And somehow that window garden made our shabby little kitchen into a special place. Even the light seemed different—more restful, more alive.

That was when I first realized I needed a garden in my life.

HOMEMADE VANILLA EXTRACT

Yes, you can do it, and it is unbelievably easy. If you place it in a pretty glass bottle, it makes a lovely little gift.

1 vanilla bean
One 4-ounce bottle with top
Scant 4 ounces vodka

Split the bean in half, put in the bottle, and pour in the vodka. Cap and let sit at least one month. (The longer, the stronger.)

WAKE-UP CALL

I was standing in our community garden looking at the colossal collection of cantaloupe that I had grown—my first-ever attempt. Vistas of an alternative career opened—I would leave my career as a physician and become a champion cantaloupe farmer! (Little did I know this was one of life's "fabulous firsts." That day was over twenty years ago and since that time I've not successfully grown even one more.)

As I stood relishing my achievement, out of the corner of my eye I spotted a lovely teenage girl in a nearby plot. She was staking up some tomato plants. She was a patient of mine, a schizophrenic. She had her head cocked, listening intently. There was a relaxed smile on her face, and I realized that in this peaceful place, she was not hearing her usual voices of terror. She was listening to the whisper of the winds and the song of the birds. I left quietly.

NATURAL HEADACHE REMEDIES

Here are two ways to handle a headache with ingredients from the garden. The first is a Midwestern pioneer recipe; the second is an old-fashioned German cure.

As an instrument of planetary home repair,
it is hard to imagine anything as safe as a tree.
JONATHAN WEINER

HEADACHE PILLOW

$\frac{1}{2}$ ounce ground cloves
2 ounces dried lavender
2 ounces dried marjoram
2 ounces dried rose petals
2 ounces detony rose leaf
1 teaspoon orris root
2 pieces of cotton batting, slightly smaller than a handkerchief
2 handkerchiefs
Lace or ribbon, optional

Grind spices, flowers, and orris root together, either by hand with a mortar and pestle or in the food processor. Pack the powder in between the two pieces of cotton. Sew together three sides of the two handkerchiefs. Place the cotton "pillow" inside and hand sew the fourth side tight enough that the contents don't leak out. Decorate with lace or ribbons, if desired. To use, lie on the pillow and inhale fragrance or place over eyes.

HEADACHE REMEDY

1 quart white rose petals
1 quart jar, sterilized
About 1 quart 90 proof vodka or rubbing alcohol

Pack the jar with the rose petals. Pour the vodka over and let stand, covered, for at least 24 hours. Rub on forehead, temples, and back of neck.

WINDOW-BOX BASICS

The key to colorful window boxes is choosing the right combination of plants. One really effective combination is a single type of flower in just one color: bright yellow begonias, for example, or all red petunias. Or you can try a variety of blue flowers: pale and dark lobelias combined with anchusa capensis is a wonderful combination. Or the tried-and-true pink geraniums with white petunias and variegated ivy. Think also about shapes—trailing combined with bushy and upright and foliage—greens break up the arrangement and make it eye-catching. Consider where the box is located and choose plants appropriately. If it is in shade all day, be sure all the plants you choose are shade loving. If one type of plant likes a lot of water and another doesn't, it would be best not to plant them together. Purchase enough plants so they can be tightly packed into the box. This will give a lush display when they flower.

Every flower about a house certifies to the refinement of somebody. Every vine climbing and blossoming tells of love and joy.
ROBERT INGERSOLL

To make a window box, make sure there is adequate drainage from the container; if not, drill a couple of small holes. Add a layer of broken terra-cotta pieces and then fill with dirt half way. Experiment with placement while plants are still in their original containers: tall, upright flowers in the back, trailing ones at front and sides. When you have a pleasing arrangement, water plants thoroughly, then remove from pots and fill dirt to ¾ inch below the rim of the box. Water thoroughly and add soil if necessary.

Be sure to water frequently, even once or twice a day during the hottest weather, and feed with a liquid plant food once a week. If frequent watering is difficult for you, consider buying a window box with a water reservoir. To keep your box looking lovely, pinch back young plants and prune leggy stems. Deadheading will help plants produce more flowers.

FRAGRANT PLANTS

Smell is so individual—I love narcissus, but know many people who can't stand it, and folks wax eloquent about wisteria, the smell of which makes me sick. So in creating a fragrant garden, let your nose be your guide. Here are some suggestions: jasmine, honeysuckle, sweet autumn clematis, mimosa, hosta, stock, evening primrose, nicotiana, angel trumpet (especially the white), moonflower, sweet pea, ginger, lily of the valley, peony, and pinks.

PERSONALIZED PUMPKINS

The work of a garden bears visible fruits—in a world where most of our labours seem suspiciously meaningless.
PAM BROWN

If you grow pumpkins, you can scratch names, dates, phrases, hearts, etc., into them with an ice pick or other sharp implement as they begin to grow. The scratch will heal over, and as the pumpkin gets bigger, so does your message.

(Just think of the possibilities—kids especially enjoy such a surprise.)

PUMPKIN SAGE SOUP

2 teaspoons olive oil

2 large onions, chopped

3 pounds pumpkin, seeded, peeled, and diced, or

1 large container canned pumpkin

3 garlic cloves, peeled and chopped

4 cups water, approximately

1 cup cooked rice

1 teaspoon fresh sage, minced

2 teaspoons salt

1 teaspoon ground white pepper

Minced parsley, for garnish

Heat olive oil in a large heavy stockpot. Add onions and cook about 5 minutes on medium/high heat. Add raw pumpkin and garlic; cook for 30 minutes, stirring frequently until pumpkin is tender. (If you are using cooked pumpkin, add with the water and rice.) Add water, rice, sage, salt, and pepper. Stir well and cook for 10 minutes to meld the flavors. Purée the soup in a food processor. Check the seasoning and consistency. Add more water if necessary. Garnish with minced parsley. Serve piping hot. Serves 6.

FRESH HERBS

One benefit of an herb patch in your garden is that you can just pick what you need for that night's dinner: no worrying about fresh parsley or basil rotting in the bottom drawer of the fridge. But even inveterate gardeners end up having to buy herbs outside of the growing season and keeping them fresh can be a real struggle. If you follow a few simple tips, however, you can greatly expand their life. The trick is to treat them as you would cut flowers.

First, untie and immerse in cool water; don't run under the faucet, that can damage tender leaves. Pick through and discard any rotting stems or leaves. Shake herbs dry gently; never use a salad spinner: it's too rough. Place the bunch stems down in a vase or canning jar that allows the leaves to stay above the rim. With basil, just store on your counter top; it will keep up to a month and may even sprout roots. With all other fresh herbs, loosely cover with a plastic bag and stick in the refrigerator. Change water every few days. Chervil, chives, dill, thyme, and watercress will keep up to a week like this; cilantro and tarragon, two weeks; and parsley as long as three weeks.

*In search of my
mother's garden,
I found my own.*
ALICE WALKER

HERB BREAD

1 teaspoon sugar

4 cups warm water

1 tablespoon yeast

12 cups bread flour

1 tablespoon salt

3 tablespoons chopped fresh basil

2 tablespoons chopped rosemary

1 cup sun-dried tomatoes, drained and chopped

$\frac{2}{3}$ cup olive oil

Extra oil and rosemary for topping

In a small bowl, combine sugar, $\frac{2}{3}$ cup water, and yeast. Let sit in a warm spot until frothy, about 10 minutes. In a large bowl, combine the flour, salt, herbs, and tomatoes. Add the oil and the yeast mixture, then gradually add the remaining warm water. As dough gets stiff, mix with your hands, until it is soft but not sticky.

Turn onto a lightly floured surface and knead for 5 minutes. Place back in the bowl, cover with a towel, and place in a warm spot until doubled in size, about 40 minutes.

Preheat oven to 425°F. Knead again until elastic, then cut into three equal pieces. Shape each into a round and arrange on oiled baking sheets. Brush a bit of oil on each loaf and top with a few rosemary leaves. Bake until golden brown and hollow sounding when tapped, about 25 minutes. Makes three 7-inch loaves.

GROW YOUR OWN

Did you know that you can grow your own loofah sponges? They are actually gourds and are available through many garden catalogs. Plant now and you can harvest next fall, enough not only for your family, but to give as gifts. To use, let the gourd ripen on the vine (it

Good gardening is very simple, really. You just have to learn to think like a plant.

BARBARA
DAMROSCH

turns from green to yellow as it ripens.) But don't let it get fully yellow—slightly green means it will be a more tender sponge. When it's time to harvest, cut off the vine, peel the skin like an orange, and let it dry for about 10 days. Then cut it open from the big end, remove the seeds by shaking and strip off any more skin. Rinse the inside fibers and then submerge the sponge in water for 12 hours. Peel off the outside layer if any remains, and dry in the shade. If the sponge is too hard, you can soften it by boiling it in water for 5 minutes.

DIG DEEP

This spring, as I dig deeper and deeper—hardened clay is not easy to break up—I'm struck by how much work that never shows overtly goes into a successful garden. It will be months before I find out if my digging is worth it. Today I'm just digging, not looking for a quick fix, adding what is needed. Ah, yes.

THE WAY TO A WOMAN'S HEART
IS THROUGH HER NOSE

I have always been extremely sensitive to smells. Blessed (or cursed) by a finely tuned sense of smell, I find I am often led around by my nose. I have fallen in love because of the way a man smelled; when I was a child and my parents were away on a trip, I used to steal into their bathroom and smell their robes hanging on the back of the door.

Innumerable as the stars of night, Or stars of morning, dewdrops which the sun Impearls on every leaf and every flower.
MILTON

One of my favorite books is *Perfume,* the story of a man so affected by scents he can smell them from hundreds of miles away.

Naturally enough, I am attracted to flowers primarily for their scent. All my roses are chosen for odor—spicy-sweet, musky, peppery—if they don't smell good, I don't want them. My current favorite is a climber called Angel Face. I also love the heady smell of lavender, the spiciness of daffodils, the romance of lilacs and lilies of the valley, and the subtlety of certain bearded irises. I particularly love the elusiveness of fragrance. You catch a scent in the garden and follow your nose to . . . where? Now it's here; then it's gone. That's why I love the sweet olive tree that blooms in southern California in early spring. The fragrance is strong in the early evening as you walk down the street, but press your nose against a blossom and the scent diminishes.

My husband, who knows of my fragrance passion, surprised me last spring by planting a huge patch of multicolored sweet peas and an entire bed of rubrum and Casablanca lilies. Batches of sweet peas perfumed my office throughout the spring. Extremely long-lasting as cut flowers, the lilies bloomed for two solid months during the summer and, all that time, the house was full of their heady scent. I don't think any gift has ever pleased me more.

ROSE WINE

Here's an old-fashioned treat. Don't do this if you spray your roses with insecticide. Be sure to thoroughly clean the roses and do not store wine in metal containers or stir with metal utensils; metal reacts to the acid in wine.

2 oranges
3 quarts rose petals, washed and lightly packed
1 gallon boiling water
3 pounds sugar
1 package yeast
5 white peppercorns

Peel the oranges and set aside. Cut up the peel. Place the rose petals in a large saucepan. Pour the boiling water in and add the orange peel and sugar. Boil for 20 minutes and remove from heat and cool. Add the yeast dissolved in warm water per package instructions, the juice from the oranges, and the peppercorns. Pour into a stoneware crock, cover and let sit where temperature is between 60°–80°F for two weeks. Strain, discard petals, rinds, and peppercorns, and bottle in sterilized jars, corking loosely for about 3 months or until the wine has completed fermenting. To store wine, seal bottles with paraffin. Makes about 1 gallon.

ATTRACTING THE BIRDS

I was at my parents' home on Cape Cod last October and became mesmerized by the chickadees, juncos, and finches that came right up to the dining room window to eat from feeders attached by suction cups to the glass. I've always had bird feeders, but they were hung from trees quite a way from the house and, so, while I had the inner satisfaction of doing a good deed, my eyes missed out on the pleasure of the darting creatures themselves. So I decided when I returned home to try and lure the birds to the window. I learned a bit in the process.

No occupation is so delightful to me as the culture of the earth . . . and no culture comparable to that of the garden. . . . But though an old man, I am but a young gardener.

THOMAS
JEFFERSON

First, the best time to start is late fall, because the cold weather makes birds more anxious to find food and therefore more willing to try something new. Birds' body temperatures are on average ten degrees warmer than humans, and they need an almost constant food supply to stay alive. The cold weather slows down and kills the insects they eat and so they must find an alternate food supply.

It doesn't matter which side of the house you put the feeders on, but it should be protected from the wind. You should have a good view, but it shouldn't be so active in the room that the birds are scared off all the time. At my parents, birds would come when we were all sitting down quietly, but every time someone moved close to the window, they would scatter. A feeder on a pole in full view of the window but back a little turned out to be the best compromise at my house. The birds felt safer than right at the window, and I could still see them (plus cats and squirrels couldn't get the birds or seed.) As for the food, wild birdseed mix is fine, as is cracked chick feed mixed with sunflower seeds for the seed-eating birds. Bug eaters such as woodpeckers shy away from seeds, but love suet; you can buy special suet feeders at any bird or nature store.

BUG-EATING BIRDS

Another way to eliminate the need for pesticides is to encourage our feathered friends to dine at your house. Here's how: (1) Place bird-baths throughout the garden, away from bushes that can hide cats. Keep them clean and filled with fresh water. (2) Create a fountain with trickling water; the sound and movement is attractive to birds. (3) Plant a diverse array of plants, particularly natives. (4) Don't use any chemicals on your garden as they can kill birds. For answers to all kinds of bird questions, you can call the National Audubon Society's Bird and Wildlife Information Center (212-979-3080; Web site www.audubon.org).

ALMOST HOMEGROWN

Flowers leave some of their fragrance in the hand that bestows them.
CHINESE PROVERB

This story isn't about my own garden; it's about someone else's. About six months ago, I saw an ad for organic produce, delivered weekly to your door. I called the number and decided to sign up. I've been delighted. The box shows up on my doorstep every Friday afternoon, and it gives me great pleasure to eat only organic foods and to help provide employment for small farmers.

Eating in season, though, has taken some getting used to—in the winter there was a time when I thought if I had to eat another eggplant I was going to go nuts and right now we're suffering from an overabundance of basil—and I still supplement occasionally by buying something I can't live without from the store. But it has made me eat things I would never have bought (spring garlic shoots are incredible!), and I find I'm eating more fruits and vegetables in general than ever before. Mostly I like the groundedness of it all—when it's spinach season, that's what you eat; when there are peaches, you can have them. And when the season is over, that's it till next year. Just another way of reminding me of the cycles of life.

PEACHES AND CREAM MOISTURIZER

Blend together one peach and enough heavy cream in a blender or a food processor to create a spreadable consistency. Massage onto your skin when needed. Refrigerate unused portion. (And use up within a day or so or it will "turn.")

THE ULTIMATE MUD PIE

The first week of January, when the rain is pouring down in buckets, I set aside a day to get out in the garden in my rubber boots and my most disreputable clothes. This is the time of year when the compost box looks its worst—a bulging mix of leaves, twigs, weeds, flower stalks, and kitchen waste, all soggy and far too cold to decay. I take the biggest fork I can lay my hands on and take the slats off the box and fork the whole mess onto the ground. Then I make separate piles of other ingredients: a dozen garbage bags of horse manure, ripe, pungent, and full of promise; a heap of sea lettuce, dragged up from the beach after the November storms; and a pile of ordinary dirt. Then comes the best part: piling it all together in layers like a really gooey Black Forest cake, except with manure instead of cherries.

Nature: The Unseen Intelligence which loved us into being, and is disposing of us by the same token.
ELBERT HUBBARD

By now, I'm covered in dirt and rain and bits of seaweed, and I'm sliding around in the muck and having the time of my life, though anybody else might think I was doing an unpleasant chore. Finally I cover the new compost heap in black plastic and head indoors for the hottest bath I can stand.

And then comes the moment of greatest pleasure: the sunny day in February when I peel back the plastic and smell the rich odor of decay and see the warm steam rising. A fork plunged into the pile reveals my greatest hope come true: a mass of seething worms, working their little hearts out to turn my compost pile into rich black soil. There's no magic like it.

HOMEGROWN IS BEST

Quite often, almost as an afterthought as I am on my way out the door, I bring a small bouquet of seasonal flowers to a friend. What is in season is humble and hardly comparable to the thing a florist would deliver to your door. But the fragrance of fresh cut sweet peas or the unexpected loveliness of onion tops and heliotrope mixed in with a few puny dahlias and zinnias make for a delightful surprise gift that is as wonderful to give as it is to receive.

WILDFLOWER POWER

I don't know about you, but I believe a lawn is vastly overrated. It takes a tremendous amount of water and causes chemicals to be dumped into our water supply. So I decided to dig mine up and plant a wildflower meadow instead. It was a glorious sight and virtually maintenance free.

The tricks are to till the soil in the spring, select a pure wildflower mix (no grass or vermiculite filler) appropriate to your growing area, and blend the seed with four times its volume of fine sand so it will disperse evenly. After you've spread it over the dirt, lay down a layer of loose hay to keep the seeds from blowing away. Usually the mixes are a combination of annuals, biannuals, and perennials. And to keep the annuals going, you have to rough up parts of the soil and reseed those every year.

May all your weeds be wildflowers.
GARDENING PLAQUE

WEED ARRANGING

You can use a wildflower meadow to create beautiful arrangements. All you need is access to a field and a bit of imagination. Wild Queen Anne's lace, bittersweet, winter cress, sea grape leaves, chive flowers, wild mustard, thistles, horsetails, and goldenrod all look wonderful in a simple vase, either all one variety or a combination. Even the simplicity of dried grasses or bare willow branches can be beautiful, while various seed pods, when dried, can make extraordinary decorations.

Remember to take only what you need, never wholly depleting even the weediest of wayside plants.

DRAGON-LADY DAHLIAS

My Aunt Myrtle had a small house where she cultivated one of the most envied gardens in northern Alabama. Her lawn was quilted with daffodils and tulips in the spring, and in summer she grew legions of leggy sunflowers and brilliant cannas. My favorites were her dahlias. They grew taller than me, and some of the flowers were bigger than my head. She would pick me a bouquet and we'd pretend they were elegant, feathery hats as I held them behind my ears. They came in all shades from lipstick pink to bright yellow, but there was one shade of red so intense it seemed to smolder. My aunt called these her dragon-lady dahlias. For me they conjured up fantasy meetings with beautiful women dressed in exotic silk gowns with beaded headdresses and long, tiny nails with crimson polish. They were intoxicating to look at, and I could spend hours engaged in my own little dramas amid their searing faces. I buy dahlia bouquets every chance I get but have yet to find any as alluring as Aunt Myrtle's dragon ladies.

And what is
so rare as a day
in June?
Then, if ever,
come perfect days;
Then Heaven tries the
earth if it be in tune,
And over it softly her
warm ear lays.
JAMES RUSSELL
LOWELL

WEATHER SIGNS

According to folklore, it will be a hard winter if

- ❀ there is an unusually large crop of acorns;
- ❀ heavy moss appears on the north side of trees;
- ❀ grape leaves turn color early in the season;
- ❀ corn has thick husks;
- ❀ hornets have triple-insulated nests;
- ❀ cattle get rough coats and rabbits and squirrels have heavier fur than usual;
- ❀ wooly bear caterpillars are black all over. If dark only in the middle, only midwinter will be hard; if the ends are darker, the beginning and end of the winter will be hard.

FAIRY BOUQUETS

In his garden every man may be his own artist without apology or explanation.
LOUISE BEEBE WILDER

My dad was a great gardener, and when I was a little girl, he used to take me out early in the summer morning before my mother woke up to make fairy bouquets. We would pick only the tiniest flowers—violets, miniature roses, baby nasturtiums—place them into tiny bud vases or old spice jars, sometimes adding a bit of ribbon around the neck of the container, and put them at our breakfast plate.

A HATFUL OF BEAUTY

For a summer garden party, beautify your old straw hat with a ribbon of fresh flowers. All you need besides the hat is raffia, scissors, and some flowers. Try flowers that will last a long time out of water—sunflowers, statice, yarrow, chrysanthemums, carnations, and daisies are best, as are sturdy greens like sword ferns. Cut the greens and flowers with long enough stems to bundle. Separate each kind and create hand-size bunches of each. Set aside. Cut several strands of raffia long enough to not only go around the hat's crown, but wrap around the flower bundles. Knot the strands together at one end. Starting 6 inches from the knotted end, twist the raffia around the stems of one bunch of flowers several times. Lay the next bunch down as far apart from the previous bunch as you like, and repeat the wrapping process. Continue down the length of raffia until you have a garland long enough to go around the hat. Place the garland on the hat and tie the two ends of raffia together; tuck any loose raffia under the flowers. To keep fresh for several days, store hat in the refrigerator in a plastic bag.

ACKNOWLEDGMENTS

A generous thank you to all who were involved in this project, from its inception to its reincarnation. A special thank you to Carolyn Rapp, whose own book of garden stories is a constant inspiration. To the wonderful staff at Red Wheel/Weiser and Conari Press: Brenda Knight, Rachel Leach, Jordan Overby, Maija Tollefson, Landon Eber, and beloved publisher Jan Johnson. Special thanks to Donna Linden for her beautiful design, and to Caroline Pincus for her eagle eye. Deep gratitude to Amber Guetebier for her artful arrangement, green thumb, and editorial magic, without which *Every Garden Is a Story* would never have happened. And an extra special thank you to the darling Rosemary Rouhana.

Thanks also to Dawna Markova, especially for her stories *The Garden Grows You* and *May, Home,* and to the past efforts of both Ame Beanland, and Will Glennon.

We also owe a debt of gratitude to all those who contributed to *Simple Pleasures of the Garden* (Conari, 1998), which was used as a sourcebook for *Every Garden Is a Story:* Mark Akins, Suzanne Albertson, The Barnstable, MA Grubs Garden Club, Jennifer Brontsema, Bill Edelstein, Julie Gleeson, Nina Lesowitz, Heather Denver, Annette Madden, Everton Lopez, Tom King, Jay Kahn, Lara Morris, Ann and Laura Marceau, Marilyn Chandler McIntyre, Patricia Renton, Gloria and Vincent Ryan, and Barb Parmet, Claudia Schaab, who provided me with original stories and suggestions. Thanks also to my husband, Don McIlraith, whose inspiration in the garden and the home have made this book possible.

Previously published contributions: Beverley Nichols in *Down the Garden Path*. New York: Doubleday, 1932.

RESOURCES

GARDEN SUPPLIES AND SUNDRIES

Antique Rose Emporium
www.antiqueroseemporium.com
1-800-441-0002
Texas based company that offers an extensive selection of heirloom and old variety roses. Retail stores, as well as mail order in both bare-root and two-gallon container sizes. Catalog available.

Botanical Interests
www.botanicalinterests.com
Beautiful, botanical illustrations grace the packages of these unusual and heirloom seeds, including a Certified Organic line. The company has a policy of no genetically modified plants or seeds. Web site offers a directory of retail outlets and online stores that carry Botanical Interests line (no direct buying from their Web site).

Gardener's Supply Company
www.gardeners.com
1-888-833-1412
As the name suggests, this company offers all the latest and greatest accessories for the avid or casual gardener, from garden to home, to seeds to tools. Great selection of composters and composting aids.

Mountain Rose Herbs
www.mountainroseherbs.com
1-800-879-3337
Bulk organic herbs and spices, teas, essential oils, and supplies such as beeswax, clays, and carrier oils for herbal crafts.

Peaceful Valley Farm Supply
www.groworganic.com
1-888-784-1722
Although they now offer a retail location in Grass Valley, CA, Peaceful Valley is best known for their print catalog, which offers seeds, plants, bulbs, and extensive garden supplies from propagation to harvest, for the serious organic gardener.

Renee's Garden
www.reneesgarden.com
1-888-880-7228
Quality seeds for heirloom and cottage garden flowers, gourmet veggies, and culinary herbs from around the world. Seed packets are beautifully illustrated. Online or mail order.

Seeds of Change
www.seedsofchange.com
1-888-762-7553
Organic seed and garden supply company that now offers a line of organic food products. Online or mail order.

Sloat Garden Center
www.sloatgardens.com
Retail garden centers throughout the San Francisco Bay Area, including Marin and Sonoma counties. Sloat offers extensive low-cost seminars to the public.

Smith and Hawken
www.smithandhawken.com
1-800-940-1170
Flowering plant subscriptions, garden gifts, statuary, high-end outdoor furniture, tools, and more. Retail and outlet stores located through out the United States.

Vintage Gardens
www.vintagegardens.com
1-707-829-2035
"Antique and extraordinary roses," they also offer heirloom and unusual hydrangea varieties. Mail order, retail location in Sebastopol, CA. Vintage Gardens offers an informative newsletter, and an extensive 360 page catalog.

White Flower Farms
www.whiteflowerfarms.com
1-800-503-9624
Premier source for mail or online orders, full-color catalog featuring plants, bulbs, and gardening supplies.

BOTANICAL GARDENS AND SOCIETIES

National Audubon Society
www.audubon.org
1-212-979-3000
Check the Web site for local chapters near you, where you will find groups and action alerts. More than just a society for bird protection, the Audubon Society has evolved into a full-fledged environmental action group.

Seed Saver's Exchange
www.seedsavers.org
1-563-382-5990
A nonprofit organization of gardeners who save and share heirloom seeds, the Web site and catalog offer an extensive selection of books, native seeds, and heirloom potato, garlic, and bean varieties.

TO OUR READERS

Conari Press, an imprint of Red Wheel/Weiser, publishes books on topics ranging from spirituality, personal growth, and relationships to women's issues, parenting, and social issues. Our mission is to publish quality books that will make a difference in people's lives—how we feel about ourselves and how we relate to one another. We value integrity, compassion, and receptivity, both in the books we publish and in the way we do business.

Our readers are our most important resource, and we value your input, suggestions, and ideas about what you would like to see published. Please feel free to contact us, to request our latest book catalog, or to be added to our mailing list.

Conari Press
An imprint of Red Wheel/Weiser, LLC
500 Third Street, Suite 230
San Francisco, CA 94107
www.redwheelweiser.com